hands off DUBLIN

by Deirdre Kelly

Photographs Pat Langan

The O'Brien Press

11 Clare Street Dublin 2

CUFFE STREET — What 'road-widening' really means!

Thank You . . .

I want to thank especially my husband Aidan for his advice and help, and my mother whose help with the children gave me much of the free time needed for the book, also Michael O'Brien who undertook to publish *Hands Off Dublin* in an unbelievably short time. Also, the Homan family of Killiney.

Thanks also to Niall Montgomery for the title, and Uinsionn MacEoin for his advice, and Mr. Sugrue of the Road Engineering Department of Dublin Corporation who was most helpful and patient on the many occasions I was studying the road plans.

The publisher acknowledges with thanks the help and co-operation of *The Irish Times* for the use and preparation of the photographic materials.

OVER TO YOU . . . .

Every citizen, man, woman and child will be affected for good or ill by the Development Plan put together and administered by Dublin Corporation. The shape of this plan which will determine the shape of your city, your neighbourhood, your street, the future of your place of work, your school, your church, your house, is the work of a team with one very important omission, *you, the citizen!*

Now is your chance to make your presence felt. Go and see this plan, study the proposals for your own area carefully. Go back a second and a third time if necessary. Call meetings and if there are any people with a specialized knowledge in this field in your area, rope them in and make use of their experience.

Ask the Planners all the questions that *they* should have asked *you* before this plan was put together. Ask for explanations. What does "zoning" mean? How will it effect the security of your home, your place of work, education, recreation and entertainment? Is road widening necessary? Is the social cost too great? Who gains and who loses?

Use your Councillors, T.D.s, the news media. Tell the Planners what you think *now – you won't have another chance for 5 years.*

The Development Plan (Draft Variations to) will be on Public Exhibition in the City Hall, Dame Street, for three months from 28th June, 1976. It will be open during normal office hours with the addition of a late opening until 9.00 p.m. every Friday.   Admission Free.

A mobile exhibition of the Plan will stop at each of the following locations for two weeks:

Pembroke Library
Charleville Mall Library
V.E.C. School, Collins Ave. Killester
Rathmines Library
V.E.C. School, Kylemore Road
      Ballyfermot
Finglas Library
Ringsend Library
Ballymun Library
Cabra Mobile Library Stop
Beaumont Mobile Library Stop
Howth Library
Raheny Library
Inchicore Library
Terenure Library

At the time of printing this leaflet the dates of the mobile exhibitions had not been finalised, but information as to these dates can be obtained by telephoning Mr. P. Byrne Phone No. 742951 Ext. 165. (Dublin Corporation)

These mobile exhibitions of the plan will be staffed during normal office hours with two late openings per week until 9.00 p.m.

Contents

First published 1976
The O'Brien Press
11 Clare Street Dublin 2.

ISBN Paperback 0 905140 04 4
ISBN Hardback 0 905140 05 2

Printed by E. & T. O'Brien Ltd. Dublin.

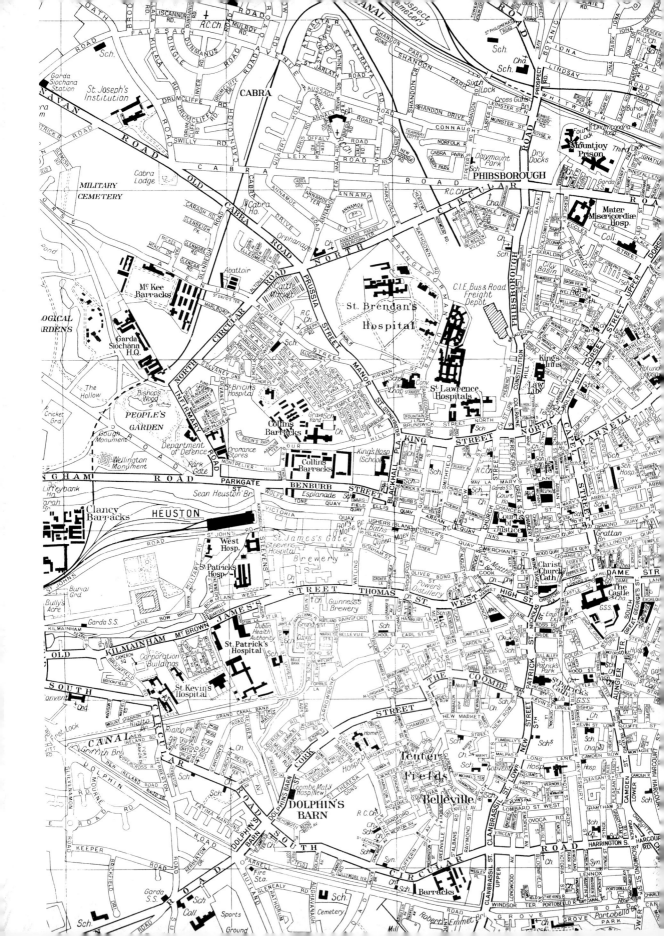

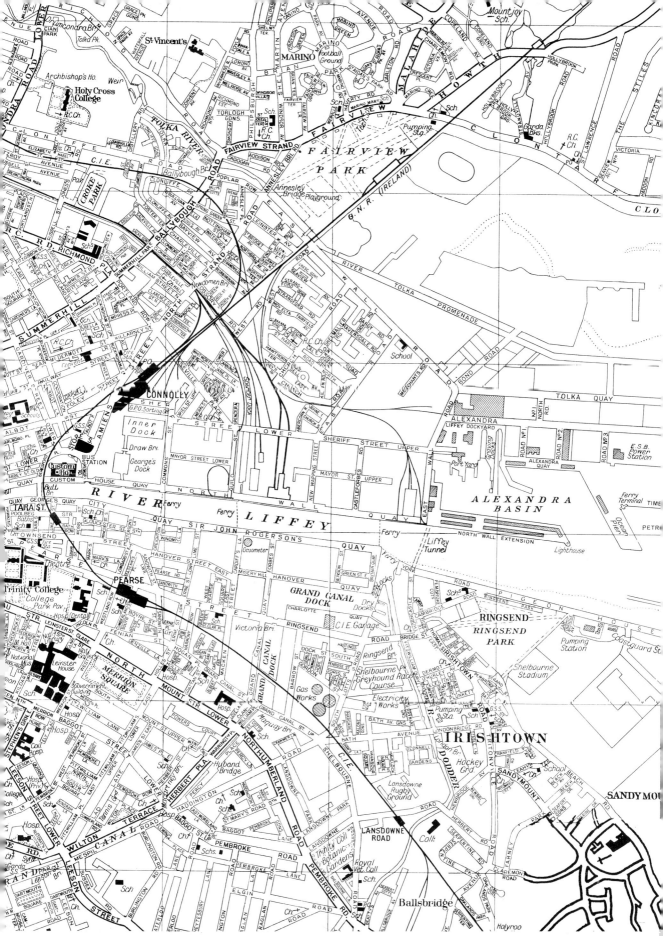

All that glitters is not . . . . .

The Corpo are at it again — a few years ago it was the Master Plan in the Black Church. The late Austin Clarke said that if you run around the Black Church twice you will see the Devil, be the janey mac the Corpo must have run around it four times, because they have the devil of a plan in City Hall. All gloss spit and polish, colours dots spots lines and dashes, all terribly well done as the Dubliner would say, a real bang on job.

Motorways for yer man and the childer — you will be there and back before ya know it. I hope yis all have your Rollis-Royces, Mercedes-Benzes and Jaguars, and when all these come into action the Corpo will Have to change the name of the city of Dublin to Cuckoo Land. The Corpo's plan reminds me of yer man who stood at the Ha'penny Bridge with his little fold up card table, playing the 'three card trick'. You remember yer man? Find-the-Lady; now it's Find-the-Motorway, or better still, find-the-people or find-the-Ancient city; instead of spot-the-ball competitions we will be able to have spot the road crossing positions.

Did you ever try to cross the road at Christchurch? A simple thing to get from Christchurch Place to the Tailors Hall, or so one would think, go up and try it, and don't think yer in luck if the lights are green, just know yer in luck if you can look left-look right at the same time and do a Ronnie Delaney sprint to the half way island. If ya get there in one piece yer in luck — and the rest of the journey is a piece of cake. But what about the old people, as well, they're old, they should not be out; I mean this is the age of Motorways man — it's go - go - go - man - go!

The Corpo should house the people in the city near their jobs, stop the office block speculators and for a change start re-building Dublin to-day for people instead of big business. The destruction of Dublin has gone on too long, it's time to call a halt. Oh and the last word, take a close look at the Railways and rid the roads of the Juggernauts and Container Trucks. It should help to cut down the unemployment figures, make the roads safer and add to the life of house accommodation and historic buildings.

Eamonn MacThomáis

Declaration of Amsterdam

Signed by all the Ministers of the Council of Europe, including Ireland, at the Congress of Amsterdam.

a)   Apart from its priceless cultural value, Europe's architectural heritage gives to her peoples the consciousness of their common history and common future. Its preservation is, therefore, a matter of vital importance.

b)   The architectural heritage includes not only individual buildings of exceptional quality and their surroundings, *but also all areas of towns or villages of historic or cultural interest.*

c)   Since these treasures are the joint possession of all the peoples of Europe, they have a joint responsibility to protect them against the growing dangers with which they are threatened — neglect and decay, deliberate demolition, incongruous new construction *and excessive traffic.*

d)   Architectural conservation must be considered, not as a marginal issue, but as a major objective of town and country planning.

e)   Local authorities, with whom most of the important planning decisions rest, have a special responsibility for the protection of the architectural heritage and should assist one another by the exchange of ideas and information.

f)   The rehabilitation of old areas should be conceived and carried out in such a way as to ensure that, where possible, this *does not necessitate a major change in the social composition of the residents.* All sections of society should share in the benefits of restoration financed by public funds.

g)   The legislative and administrative measures required should be strengthened and made more effective in all countries.

h)   To help meet the cost of restoration, adaptation and maintenance of buildings and areas of architectural or historic interest, adequate financial assistance should be made available to local authorities and financial support and fiscal relief should likewise be made available to private owners.

i)   The architectural heritage will survive only if it is appreciated by the public and in particular by the younger generation. Educational programmes for all ages should, therefore, give increased attention to this subject.

j)   Encouragement should be given to independent organisations — international, national and local — which help to awaken public interest.

k)   Since the new buildings of today will be the heritage of tomorrow, every effort must be made to ensure that contemporary architecture is of a high quality.

SPENCER DOCK

At this point it is proposed that the motorway would rise at least to the roof level of the Sheriff Street flats on the left of the photo, and follow the route of the Royal Canal. This would mean covering in the Canal.

The alternative is to plant and develop the banks as an amenity for the people living in the area. They are desperately in need of recreational space.

CLONMORE TERRACE

Here beside the Royal Canal the houses nearest the bridge would have to go to make room for the motorway and the junction with Summerhill Parade. The other houses if not actually demolished would be "unlivable in" because of the noise.

'Urban motorways should be aligned at least 500 feet from nearest house to achieve acceptable noise levels *inside* the house.' *Motorways in the Urban Environment, British Road Federation 1971.*

In the distance on the right Charleville Mall Flats are also touching the proposed motorway route.

LOOKING TOWARDS CROKE PARK

The motorway will branch off to the right just before the sports ground.

8

ROYAL CANAL at Whitworth Road

PROSPECT ROAD AND HARTS CORNER. The buildings in the foreground of the above photo will also disappear. (See note below on Harts Corner)

HARTS CORNER

HARTS CORNER

Harts Corner is one of the best known landmarks in Dublin. A new road is to be built behind it, from Finglas Road through Prospect Avenue to Botanic Road. This new road will involve the demolition of a number of houses, and on Botanic Road the road-widening will take all or part of the gardens from 40-52. Harts Corner and the corner beside will become a large traffic island.

top left — BICYCLE SHOP AT CROSS GUNS BRIDGE. One of the delights of Dublin are the little shops such as this, which look as if they had 'grown' on the spot. This shop, like many others of its type in the city, will go if the motorway is built.

bottom left — WHITWORTH ROAD from Cross Guns Bridge. A view of part of the proposed route of the motorway along the Royal Canal.

SANTRY.

This photo shows lands reserved for a motorway at Santry. This would be the minimum amount of space needed for a motorway, and does not include the space needed for interchanges, ramps etc.

Imagine the effect of a swathe of concrete this width running along Sandymount strand, through Ringsend, past Sheriff Street flats, across recreational areas such as the sportsfield at Croke Park and through the heavily built up and populated areas along the Royal Canal, not to mention the fact that the Canal and railway line alongside it will be filled in. It will at one stroke eliminate two of Dublin's finest recreational areas, the Canal and Sandymount Strand, render a large area of the city unlivable in, and remove two alternative modes of transport.

ENTRY TO ROYAL CANAL The motorway
breaking through the buildings across the
river, will follow the route of the Royal canal
from this point on through the city.

LIFFEY CROSSING showing where it will
cross just above the gasometer.

WEST ROAD

An ironic sign at West Road

ROYAL CANAL. Boys playing on the banks of the Royal Canal behind St. Ignatius Road. Instead of becoming a motorway, this is an ideal location for the linear park, which has been proposed by many sections of the community. A strange contradiction is that Dublin Corporation has been planting trees and assisting in cleaning up the canal despite their future plans for it.

EST ROAD the motorway proposals are wavering back and forth over this road, but which er way it goes, a large section of the road will go. Three photos on left show sections which ll almost certainly disappear. The rest will be within the noise danger area and will probably ve to be vacated.

RICHMOND CRESCENT. These houses will
be swept away in the path of the motorway
if it is built.

RICHMOND CRESCENT

SUMMERHILL PARADE. Part of this street on both sides will have to go, and it is quite likely that the whole terrace will be demolished on the left of the photograph, as it will be well inside the noise danger area.
SACKVILLE GARDENS. Another charming little terrace in the path of the proposed motorway.

PARNELL STREET AND SUMMERHILL

Extract from "Conclusions of the OECD Conference in Paris — Better Towns with Less Traffic".

PARNELL STREET. If the road widening is completed in this street and in Summerhill, the way will be cleared for four to six lanes of fast moving traffic. Apart from the terrible physical and social dangers to the hundreds of families living in Summerhill, building this highway will also involve taking at least the first and second houses on both sides of all the streets adjoining or crossing them. Rutland Street, Gardiner Street and Hill Street are examples of streets where this will occur.

" These problems have led to a change of emphasis in traffic and transport policies in many cities and countries. Rather than concentrating on the accommodation of private cars, the aims are: to restore the city to a human scale and to preserve it as a centre of economic, social and cultural life. To this end actions are being taken

- to ensure or improve accessibility to activities for both people and goods;

- to reduce air pollution, noise, acci-
- dents, and other adverse effects of motor traffic;
- and to conserve energy resources. "

| ROADSPACE (a) | | |

POLLUTION: CARBON
MONOXIDE (b)

```
*                   * * * * * * * * * * * * * * * * * * * * * * * * * * * * *
                    * * * * * * * * * * * * * * * * * * * * * * * * * * * * *
                    * * * * * * * * * * * * * * * * * * * * * * * * * * * * *
                    * * * * * * * * * * * * * * * * * * * * * * * * * * * * *
                    * * * * * * * * * * * * * * * * * * * * *
```

POLLUTION: LEAD (c)   Zero   2,000 tonnes per annum used in UK petrols

FUEL (d)   PPPPPPP   PPPPPPPPPPPPP

OPERATING COSTS (e)   ££££££££   £££££££££££££££
                           ££££££   ££££££££££

Notes

(a) 1 bus = 20 cars = 10 × roadspace, from ref. **1**.

(b) Petrol vehicles 56 tonnes/million vehicle miles, and diesel vehicles 8 tonnes/million
     vehicle miles (rough estimation from refs. **2** and **3**. Assume 1 bus = 20 cars.

(c) Ref. **4**.

(d) Direct energy in kcal per passenger mile, ratio 350:630, from ref. **5**.

(e) Bus costs per vehicle mile = 26.97p, from ref. **6**.
     Typical car costs ($1\frac{1}{2}$ to 2 litre car) per vehicle mile = 2.488p (running cost only), from
     ref. 52. Assume 1 bus = 20 cars.

1. Reynolds D. J. "Economics, Town Planning and Traffic" (Institute of Economic Affairs, 1966)
2. Clean Air (Summer 1972)19
3. 'Highway Statistics' (HMSO)
4. Clean Air (Spring 1972) 13
5. 'Tyneside Passenger Transport Authority and Executive' (Annual Report and Accounts, 1971)
6. 'The AA Car Buyers Guide' 1972.
Diagram shown above is taken from 'Motorways & Transport Planning in Newcastle upon Tyne'

STONEYBATTER

MANOR STREET

PRUSSIA STREET

Stoneybatter Manor Street and Prussia
Street have a quality more like a country
town than a city. The present heavy traffic
running through these streets is destroying
this quality, particularly in Prussia Street,
one of the oldest streets in the city. This
street is narrow and totally unsuited to the
heavy through traffic to which it is
subjected. When road widening in this area
takes its toll, every bend and projection in
Stoneybatter and Manor Street will be ironed
out, and Prussia Street from Manor Street to
N.C.R. will be widened on the east side.

JUNCTION OF CHURCH STREET AND NORTH KING STREET. One of the biggest junctions on the inner city highway will built at the point shown in this photograph.

COOPERS, Horse Dealers of Queen Street

N. King Street Junction looking towards Church Street.

N. King Street junction looking towards Constitution Hill (more corporation flats on left of photo).

" Traffic fumes contain a number of air pollutants, particularly carbon monoxide and unburnt fuel (resulting from incomplete petrol combustion), oxides of nitrogen (formed in the combustion chamber from nitrogen and oxygen in the air) and lead salts (chiefly lead bromochloride; the lead originates from antiknock fuel additives, e.g. tetraethyl lead).

Potentially far more dangerous is lead, because its effects are cumulative. Inorganic lead is stored in the bones, and may be released at a much later date to produce anaemia, kidney disease and damage to the nervous system — with such effects as convulsions, brain haemorrhage and mental retardation. Tetraethyl lead, which forms a small percentage of the lead released in traffic fumes, is a psychotropic poison whose symptoms can mimic those of a psychotic or psychoneurotic disorder. Although the health hazard from present environmental lead levels is a matter of some controversy, "one must remember that lead absorbed in seemingly harmless trace quantities over a long period of time can accumulate to exceed the threshold level for potential poisoning and produce delayed toxic effects".

Atmospheric lead absorption approaches — or may, in some cases, exceed — absorption from dietary sources. As much as 98% of identifiable airborne lead may originate from cars and the level of atmospheric lead varies directly with the volume of traffic. This being the case, an urban motorway system generating large volumes of extra traffic, must present a magnified threat to public health. Lead levels in the blood of residents living near the M 6 "Spaghetti Junction" interchange have increased by 25% between May 1972 and March 1973. "

Extract from "Lead Pollution and Poisoning" by S.K. Hall, 'Environmental Science & Technology',

SUSANVILLE ROAD. The motorway will go straight through this pleasant street, demolishing most of the houses.

CLONLIFFE ROAD. Numbers 119 to 135 and 124 to 150 on this road are certainly doomed if the motorway crosses here as planned.

CROKE PARK (sportsfield)

Open Spaces

The photograph above is an example of one of the many sportsfields and open spaces this motorway will traverse. It goes straight through the sportsfield behind Croke Park. Another sportsfield which will lose a section to a new road is the Vocational Sportsfield in Terenure which will be cut at one corner by a new road going from Templeogue Road to Terenure Road West. Pearse Square (photo page 43) is another essential playground and recreational area which will go. The pleasant green crescent on Davitt Road on the Grand Canal will also lose quite a large slice. These open spaces are essential in a rapidly expanding city and it is a criminal act to take them either wholly or in part for a minority use, and thus deprive people who do not or cannot drive cars (the elderly, the very young and the infirm) of places in which they can relax.

CORNER OF GRACEPARK TERRACE AND ROAD. Underpass here will mean demolition for these houses and possibly others.

GRIFFITH AVENUE which will also have an underpass at about this point

TRINITY COLLEGE RAILINGS from the corner of Grafton Street to Nassau Street
will be pushed back 20ft, breaking the great line of massive railings along Nassau Street.

ELDERLY WOMAN TRYING TO CROSS ROAD. A feat which is almost
impossible for the elderly and infirm in our city.

DUBLIN'S QUAYS

Aesthetically, one of the most serious threats to Dublin, if the road widening goes ahead, is the destruction of the line of the quays. Both sides are to be widened — and any future developments will be set back from the existing building line. This will effectively destroy the townscape, in the same way that this policy has destroyed so many of our streetscapes. The *Architectural Review* in its special edition on Dublin made the following comment:

'Without question, it is the quays which give topographical coherence to Dublin. They are the frontispiece to the city and the nation: grand, yet human in scale, varied yet orderly, they present a picture of a satisfactory city community: it is as though two ranks of people were lined up, mildly varying in their gifts, appearance and fortune, but happily agreed on basic values.* The quays are in a bad way, with ugly gaps here and there and many buildings which seem verging on final dissolution; but they remain an immensely evocative and successful piece of townscape and the success of any move to restore Dublin may fairly be measured by whether or not it brings to the quays a return to prosperity and coherence."

PARLIAMENT STREET AND DAME STREET. All buildings shown in this photograph will be eventually knocked down to make way for a wider road. Ironically, the front section of the Olympia Theatre is included in this though only recently the theatre was granted £100,000 by Dublin Corporation *for restoration!*

WESTLAND ROW. Thirty years ago this Parish had a population of 35,000. It has now only 4,000 people. This long terrace of sound houses will be domolished for road-widening. Converted into residential units for Trinity College (the owners of most of the houses) they would bring some much needed life back into the area.

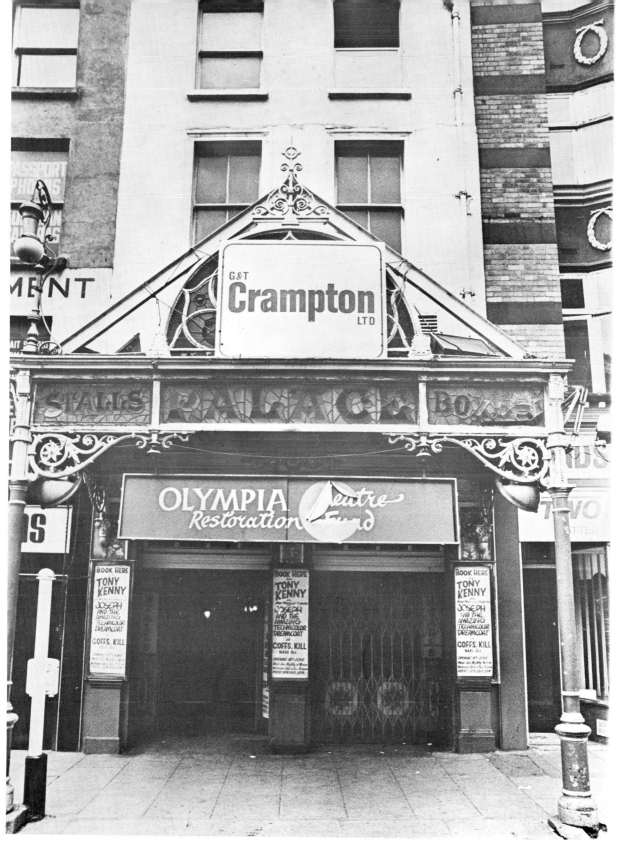

OLYMPIA THEATRE

THESE HOUSES IN PIMLICO TO-DAY —

COULD LOOK LIKE THESE HOUSES IN CORK STREET TO-MORROW

One could follow the route of any major road widening scheme by following the rows of boarded up empty or derelict houses, gap toothed terraces, and ugly rat infested sites which make the centre of Dublin look like a bombed city. It's not to-day or yesterday that these C.P.O.s were imposed. People in parts of the city have had this threat hang - ing over their homes for thirty or forty years.

Despite a steadily worsening housing situation, under successive governments more and more houses were put under C.P.O. and allowed to decay to make way? for the 'ubran motorway system.

Indeed in retrospect is is not strange that no government or city council ever made any serious attempt to stop the deliberate vandal- ism of houses by unscrupulous developers, despite repeated calls on them to do so. How could they when everywhere in the city they were committing the same crime themselves!

Some of the photos in this book are of street which are threatened by 'long term' road plans. Don't be mislead by this, twenty or thirty years ago the rows of dereliction, now in the city, were sound houses — and 'long term' road plans!

ROAD WIDENING

Sir,—Your environment corres- pondent, Dick Grogan, writing (April 13th) on future road pro- posals for Dublin, lists 120 streets "due for widening." The list appears under a heading which reads "The five-year forecast."

This is a complete misrepresen- tation of the factual position and is a contradiction of the reference to the same 120 streets in the main article which states "those streets on which works are planned some time *after* the next five years."

The streets in question are being earmarked for "long term roads reservations", i.e. if redevelopment takes place on these streets in future years a set-back of build- ings *may* be required. What this simply means is that Dublin Cor- poration as the roads authority for Dublin city is carrying out its legislative responsibility in plan- ning for the prevention of develop- ments that may inhibit possible road proposals in the distant future. Not to make such long-term provisions would be less than re- sponsible on the part of the Cor- poration.

To interpret the issuing of a list of streets involved in this exercise as a plan for wholesale road widening in the city area is making undue certainty of something that is not certain and could cause un- necessary public concern.—Yours, etc.,

NOEL CARROLL.
Public Relations Officer,
Dublin Corporation.
City Hall,
Dublbn 2.

THOMAS COURT — to disappear in road widening from School Street to Thomas Street.

FRANCIS STREET

Francis Street — vibrant with life and char -
acter — will be widened from the Iveagh
Market to Cornmarket and will become part
of the great junction at Christchurch place
and High Street.

This intimate street with its shops, houses
and famous market should be renewed where
neccessary, and the street scale retained and
traffic strictly controlled. the demolition of
any houses in this area, which is close to the
city centre, will almost inevitably result in an
influx of offices when needs of the road's
department have been satisfied.

CORNER LANDMARKS

Many crossroads in Dublin are characterised by 'corner shops' or pubs. This part-icular cross at the junction of Kevin Street, Patrick Street, Dean Street and New Street, had a pub on each corner. One is now demolished, and when the road widening is completed here every building in these photos will be flattened and the sites will become part of an enormous junction.

Dead Corner

Living Corner

ST. LUKE'
WITHOUT
This beauti
old church
with its alm
houses
will have a
road runni
between it a
the almshou
It is very po
sible that th
almshouses
will have to
demolished

PATRICK STREET AND CATHEDRAL

This Cathedral built about 1225 will have a major dual carriageway passing its doors. Most of the sur-rounding streets will be demolished for road widen-ing and Patrick Street itself will be flattened from end to end on the east side (right on photograph).

CHRIST CHURCH built in 12th century a new road from Winetavern Street to High Street will emerge at this point (see map over-leaf) Dereliction, as usual, marking the point.

ST AUDEONS on the left of the photo is the oldest parish church in Dublin, built from 1190 onwards. The vibrations and fumes from heavy traffic could do irreparable damage to this and other ancient buildings.

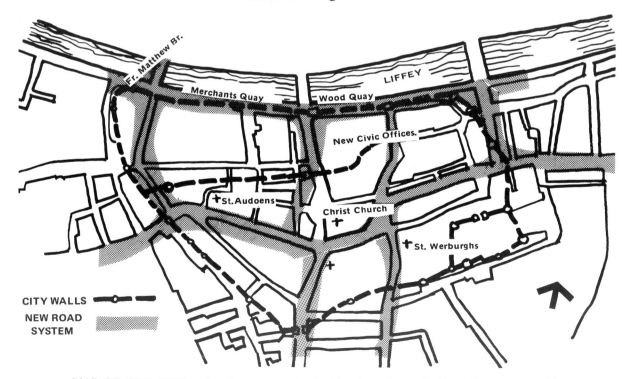

MAP OF OLD CITY showing new roads. Dublin City Council's Road Programme will raze the city within it's walls.

CORK STREET. The building facing down Cork Street in this photograph will be demolished, as the new road (from New Street to Ardee Street) will emerge at this point to continue through Cork Street which will be widened for its entire length on the South side.

CHRISTCHURCH PLACE
AND HIGH STREET

The area within the walls of the old city (see map on left) rich in history, an archaeological area of international importance and once the most heavily populated part of the city, seems to have been singled out by Dublin Corporation for utter devastation. Having depopulated large sections, the planners are carrying out a programme of major new roads, junctions and road widening which will turn this history steeped living entity into a criss-cross network of roadways.

Beside Christ Church on one of the most important archaeological sites in Europe, Dublin Corporation chose to build its new civic offices; a complex of huge office blocks, completely out of sympathy with its sur-roundings. The deep excavation required for the foundations would have wiped out the evidence of centuries of habitation going back to the origins of the city. To fit in with the building programme the time given for archaeological investigation was cut to a minimum, forcing archaeologists to race neck and neck with the bulldozers and leading to a situation where pieces of the mediaeval city were found in the city dump at Ringsend.

Recently it has been decided to postpone work on the new civic offices due to lack of funds. The major road network may never be completed for the same reason, but irrevocable damage will have been done both historically (not only nationally but in a European context) and socially.

DOLPHIN'S BARN STREET

CUFFE STREET. Until recently a narrow road capable of taking only two lanes of traffic at one end. The corporations method of relieving this 'bottle neck' can be seen on the opposite page. However the fast moving traffic which can now zoom into Cuffe Street will meet another bottle neck at Kevin Street (bottom photo), the corporation have prepared for that and intend to demolish one side of that street to clear that bottle neck and so on until there's no bottle necks or sharp turns on the streets of Dublin. Then there will be no Dublin.

DOLPHIN'S BARN STREET. A narrow winding street very characteristic of Old Dublin, with sound houses all along one side and empty spaces where houses once stood (until c.p.o'd for road widening) along the other. A major arterial road will make the remaining houses unlivable in with noise and pollution.

'TIME FOR A CHAT AT THE CORNER SHOP'. But time is running out for the little corner shops, and who can chat at the edge of a dual carriageway?

OLD DOORWAY in Pimlico, under road widening threat.

PEARSE SQUARE. The motorway raised at this point, will destroy the houses and also the square, which is the only recreation space in this area.

SANDYMOUNT TOWER

SANDYMOUNT STRAND

44

RECLAIMED LAND at Sandymount.

SANDYMOUNT STRAND

Sandymount Strand, unique in its closeness to the city centre. Immortalised by Joyce, and part of the childhood of thousands of Dublin people, this strand is under constant attack. The same city council which declared Dublin Bay "An area of Special Amenity" and which opposed the building of an oil refinery in the Bay, have to decide whether the motorway will cross the strand from Booterstown to Ringsend, making a farce of any amenity order.

The Port and Docks Board have plans for filling in 2000 acres of the strand and this reclamation will only be a matter of time if this motorway is built.

In recent years large sections of the strand have been reclaimed by dumping. Some of this has been used for industry. Also, huge drainage pipes were laid along the foreshore and the fine white sand was replaced by the polite promenade shown in the photo-promenade is laid out with fine car parks (again pandering to the motorcar) but what good is that to a mother with several children who can only come by bus?

It should be noted that most of the above was done without any public consultation with the thousands of people who use the strand.

While recognising the urgent need for jobs and industry, it is difficult to understand why factories and industries are being moved out of the city to industrial estates, without a bleep out of anybody concerned, yet at the same time an irreplacable amenity is being destroyed to give land for industry and factories in the city. It's a strange world.

COALYARD in Ringsend
which will go in the huge
junction at this point.

FACTORY at York Road in
the path of the motorway.

COALYARD in Ringsend
which will go in the huge
junction at this point.

GRAND CANAL HARBOUR. This could be a wonderful amenity, as a marina for the Grand Canal, bringing life and colour into this area, but not if the canal link to the motorway comes down the centre as is proposed in the Road Plans.

CANAL BANK WALK

Leafy-with-Love banks and the green waters of the canal
Pouring redemption for me, that I do
The will of God, wallow in the habitual, the banal,
Grow with nature again as before I grew.
The bright stick trapped, the breeze adding a third
Party to the couple kissing on an old seat,
And a bird gathering materials for the nest for the Word
Eloquently new and abandoned to its delirious beat.
O unworn world enrapture me, encapture me in a web
of fabulous grass and eternal voices by a beech,
Feed the gaping need of my senses, give me ad lib
To pray unselfconsciously with overflowing speech
For this soul needs to be honoured with a new dress woven
From green and blue things and arguements that cannot be proven.

Patrick Kavanagh

ALBERT COURT   These charming cottages are doomed if a proposed new road is built from Grand Canal Street to Lower Mount Street.

GRAND CANAL STREET

Another familiar piece of Dublin streetscape, the houses on the right beyond the pub will also disappear if the above road is built.

PORTOBELLO HOUSE

Recently beautifully restored, will become a traffic island if the corporation road plans (Illustrated opposite) go ahead.

FATIMA MANSIONS

These blocks of flats have an enviable recreational area of water and open space at their doorstep but, the canal is to be narrowed and the grassy banks 'developed' into a major traffic route, with all the noise, fumes and danger which this entails.

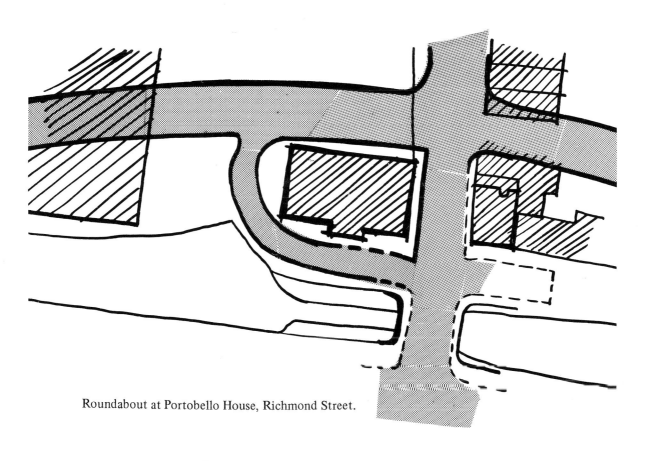

Roundabout at Portobello House, Richmond Street.

CHARLEMONT STREET AND PLACE

Most of the buildings in this photograph (including the pub) will be demolished to clear the way for a large traffic junction. The east side of Charlemont Street (right side of photograph) will also be demolished for almost its entire length.

WINDSOR TERRACE

"Rationalisation" for the new road will wipe out part of the terrace, the trees, the grass verge and all the gardens, also the plots in the photo below.

VEGETABLE PLOTS

RATIONALISATION OF THE GRAND CANAL

THE COUNTRY IN THE CITY

Part of the charm of Dublin is this rural quality seen in pockets all over the city, but particularly alongside the two canals. These fully cultivated vegetable plots show the result of years of loving husbandry. Here the use of canal banks for the production of food is something that calls for admiration and should be valued by our city fathers. Instead, this life giving soil will be buried in concrete.

Historic canal to re-open

The 200-year-old Caldon Canal, which supplies water to the main line of the Trent and Mersey canal is to be reopened to navigation on Saturday after many years of semi-dereliction.

Restoration of the canal, first opened in 1779, has taken 2½ years, half the time originally planned.

MAN WALKING HIS DOG on the Grand Canal

PORTOBELLO ROAD. This delightful little terrace of houses will go for "progress"

A lengthy battle to save the canal from being filled in to become a six lane motorway, was fought, and apparently won, some years ago. But the Dublin Transportation Study, recommended that an arterial road should run along both sides of the canal to eventually link up with a motorway at Ringsend.

If these proposals go ahead, many houses on the north bank of the canal will be demolished, the canal will be 'rationalised' i.e. narrowed to accomodate the road, (and it will be almost inaccessible at many points). The peace and tranquility of the grassy banks will be gone forever and the waterway stripped of its amenity value, will become merely a central division to a dual carriageway.

ADELAIDE ROAD

A tunnel is proposed to pass under these houses to bring the arterial road under Leeson Street Bridge will mean the destruction of this entire terrace. (See diagram).

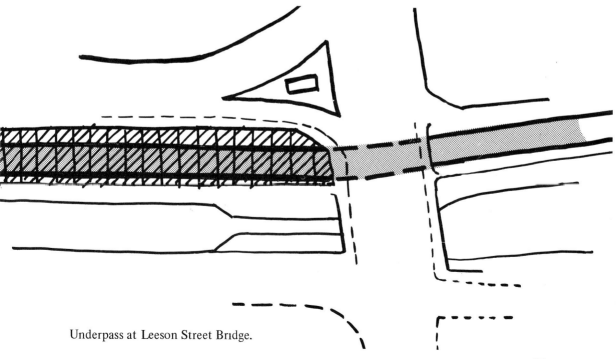

Underpass at Leeson Street Bridge.

PORTOBELLO ROAD — near Harold Cross

All along it's route this link road will cut communities off from the canal banks which have been their main recreational area for generations.

CANAL OPPOSITE PARNELL ROAD

The two canals, the Royal and the Grand, are essential lungs of water and greenery coming into the city. They cannot be replaced.

Conversation — PORTOBELLO ROAD

In the photo above, the two women chatting would find this virtually impossible in many of the streets of Dublin to-day because of the noise from traffic. These houses are doomed to go.

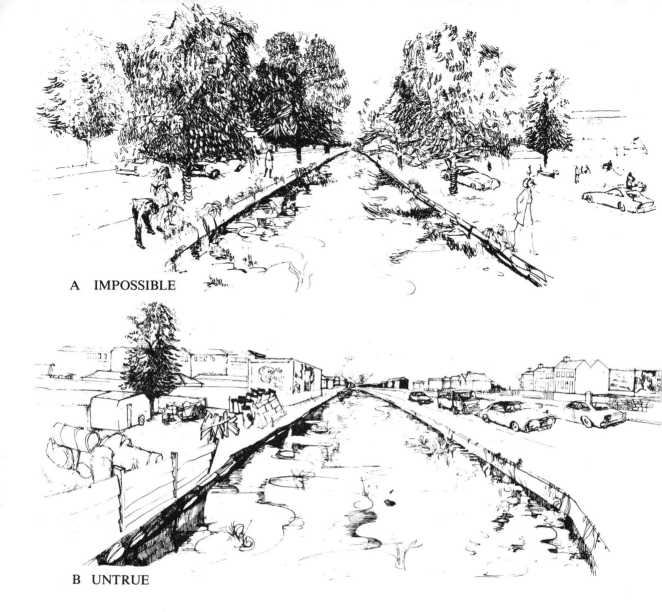

A IMPOSSIBLE

B UNTRUE

Drawing A. Taken from the D.T.S. is an artistic impression of the park-type setting envisaged for the arterial roads on either side of the Grand Canal. This is an example of the type of sugar coating which covers the bitter pill in many government authorised transport studies.

The photographs on previous pages and a study of the engineering drawings in the corporation offices show that such treatment will not be possible. The canal in fact will have to be narrowed in many places to accommodate the new roadway. It is far more likely that drawing B (taken from the same report) is what the final result will look like. Ironically drawing B was done to show how it is claimed the canal looks at present.

The report also states that: "These proposals are possible however *only* if the other sections of the *regional system* are built. The regional motorway routes, strategically placed so as to distribute long distance and heavy commercial traffic, provide the basis for preserving such amenities as the Grand Canal."

The roads on both sides of the canal are intended to link up with the regional motorway system.

It is likely that due to lack of money, or change of policy, the motorway system will not be built in the foreseeable future — if ever — and the Grand Canal 'parkway' will be commandeered for anti-social through traffic and the juggernaut.

LEESON STREET BRIDGE. This is one of the few stretches of canal which would have the width necessary to take the proposed roads. It is also one of the most beautiful parts, but already is almost inaccessible on one side because of traffic and has always a ribbon of parked cars along the other.

WOMAN ON PORTOBELLO ROAD.

BALLYFERMOT — New Road going into Ballyfermot, it will use up the small amount of open space left in this area for recreational purposes.

OLD PINE TREE PUB

This pub, a link between the old and new communities in Ballyfermot will be demolished to allow the new road (above) to pass through.

FLATS at Kevin Street

FLATS at Summerhill

FLATS at Pimlico

WOMAN WITH BIRD in Bridgefoot Street Flats

Do the Trade Unions Care?

Something which would make a great difference in planning in the city, would be if the trade unions started to take an interest in what is happening to the environment. Most of the poorer communities n the city centre are under constant threat from motorways, office blocks, traffic pollution and countless other undesirable elements. The unions are fighting for jobs, better working conditions etc. but of what benefit will this be to the average worker if the environment (which has been enjoyed to the full by the wealthy up to now) has been destroyed when they finally have the money and the leisure to enjoy it.

In other countries trade unions are begin - ning to realise this, particularly in Australia, where a system called "green bans" operate, in which in many cases unions and city residents joined together to fight undesirable developments which threatened the environ - ment. Mr. Jack Munday, secretary of the New South Wales Builders' Labourers' federation, claims that these 'green bans' have weakened the common belief that concern for the environment is the preserve of the middle and upper classes. "Workers have to live in the worst environments, so it's obvious why trade unions should care about the environment," he maintains.

For too long small groups and associations have been fighting speculation and bad plan -ning without any help from the trade union movement, which is the one group of people which could be really effective, and it is about time it started to be so.

FLATS at Bridgefoot Street

The tree planting here will be needed as a buffer from the noise of heavy traffic when the road is fully widened. The stream of traffic will also effectively sever communal relationship between the new flats on the left and the older ones on the right, (Oliver Bond Flats).

Cuffe Street

PEOPLE BEFORE TRAFFIC

All along the route of the motorway and road widening can be seen groups, large and small, or corporation flats, which will suffer extensively from the noise and pollution of the heavy traffic for which these roads are being made. Residents of these flats are generally unaware that such a road is even proposed. Usually when it is discovered, it is too late to use any legal rights which they may have. The Corporation officials say that these roads are being created to channel the traffic from the 'residential areas'. All the blocks of flats shown in the photos in this section, will be seriously effected by heavy traffic going past their doors and windows, and polluting the air they breathe. Do these flats not count as 'residential areas'.

Cuffe Street community will also be severed by a 90ft. wide dual carriageway (see photo page 41). In the area of Cuffe Street there are thousands of people living in flats with little or no recreational space. The new road has not only brought with it all the dangers of heavy traffic, but it has actually taken some of the meagre open space from the residents of the flats. The photos illustrate the dreadful closeness of the dual carriageway to the flats.

63

FLATS at Sheriff Street

FLATS at Dolphin's Barn

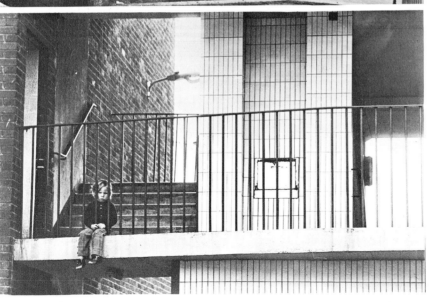

NEW FLATS at Summerhill

64

People, Traffic and Survival

Perhaps there is no better way to introduce this section than to quote from the European Architectural Heritage Year publication *The Invading Motor Car*.

"No excuse is needed for devoting a whole issue of this magazine to the motor car. Motor traffic constitutes one of the severest threats to historic buildings and to the historic character of old towns and villages. It is a threat that takes many forms. The motor car provides civic authorities with a reason for wholesale street-widening or for carving new streets through historic areas, thus destroying at a stroke both old buildings and their settings; it demands multi-lane overpasses as it enters big cities, allowing it to roar past the upper windows of residential buildings; it crowds old city centres so as to make them no longer enjoyable to explore and so as to compel us to look at famous buildings over a foreground of parked vehicles; it makes dangerous race-tracks out of narrow village streets and even — when heavy lorries fight their way through city streets — damages old buildings by their weight and impact; it can damage them also by the vibrations and air pollution it causes".

The following pages show some of the problems and dangers to our communities in Dublin and to the fabric of our city by road traffic. Also shown are some of the measures taken by various cities in other parts of the world in a new approach to transportation.

'Decrease of life' factor table, produced as a result of extensive field tests, giving the reduction of expected life in well-constructed dwellings subjected to various traffic densities. (Produced by the Research Institute for Building and Architecture of Czechoslavakia)

REDUCTION IN LIFE OF HOUSES DUE TO TRAFFIC		
Expected Life Loss	Vehicle flow Units per 24 Hours	Decrease Factor
Zero	Up to 260	1
4.0	260 to 600	2
7.0	600 to 960	3
10.0	960 to 1540	4
15.0	1540 to 2660	5
20.0	2660 to 3440	6
25.0	3440 to 4660	7
35.0	4660 to 7440	8
50.0	7440 and more	9

Extract 'European Heritage' issue two, "The Invading Motor Car", published as a contribution to the European Architectural Heritage Campaign, initiated by the Council of Europe.

TRUCK SMASHED INTO COUPLE'S HOME

By BRENDAN FARRELLY

A YOUNG married couple and their four-month-old baby were shocked today when a heavy truck smashed into their cottage home at the end of the Finglas dual carriageway leading into Dublin.

Mr. and Mrs. Noel Colley and their baby son Eoghan were asleep when the traction section of an articulated truck hit their cottage home at Tolka View Terrace. No one was injured.

The side of the cottage, facing the end of the carriageway, was reduced to rubble. A gaping hole showed through into the living room. Bricks and cement were stewn on a sofa and on the baby's pram.

Mr. Colley said they had been living in fear of such an accident. He said: "We were asleep when we heard this tremendous bang. I though a gas main had exploded".

"Now it looks as if the whole house will have to come down" said Mr. Colley a self-employed electrician.

WFIFE SHOCKED

He said his 23-year-old wife Linda was deeply shocked when she saw the rubble piled inside the pram. She was being looked after by neighbours today.

The truck, owned by J. Meredith and Sons, licensed hauliers and machinery installers, careered off the carriageway and cut a lamp-post clean off at the base.

Mr. Tholmas Meredith, of Casement Road, Finglas West, the driver of the truck, escaped injury.

There have been serveral accidents at the scene, the last two months ago.

The truck which crashed into the Molloy Cottage near the dual carriageway at Finglas.

Trains must go on tracks because of their length and to control their passage, yet juggernauts the length of railway carriages are roaming our city streets at will. Apart from serious accidents such as these and many others which have occured recently, everytime one of these monsters turns a corner in a city or town they have to mount the pavement with the resulting danger to the pedestrian and damage to the pavement. Alternatively, the corner is widened for their benefit, a task which usually involves the demolition of several houses and the loss of another piece of townscape.

BARROW STREET RINGSEND. One of the m
accidents resulting from heavy, unrestricted tra
in this area.

UCD campus to be big traffic circus?

E threat of becoming Dublin's biggest round about—with new motorways on all sides and e even cutting through the grounds—now hangs over the pleasant parkland campus of versity College, Dublin.

UNIVERSITY COLLEGE DUBLIN moved to Belfield some years ago because of alleged lack of space to expand the college in the city. Look what's happening to their new space!

The number of full-time students at U.C.D. may have grown, but the campus will shrink further if current road building plans go through. Already the widening of the Stillorgan road has cost the college 14 acres.

Now, Dublin Co. Council have plans for a motorway along Foster Ave. and a dual-carriageway connecting with the Enniskerry road which will affect a further 32½ acres of the Belfield campus.

Twelve acres are reserved at the Foster Ave. side of the campus for the proposed motorway but it appears that a further acre and a half will also be lost.

Reserved

The amount of land reserved for the Enniskerry trunk road is six acres, but a further 13 acres will also have to go to accommodate the project. So will a boiler house and water tower.

The college has appealed against the plan which, it is claimed, would spoil the integrity of the campus and interfere with the building programme for playing fields, faculty buildings, residences and car parking.

67

Towns are better with less traffic: so long as adequate provision is made for the mobility of workers and residents, and the distribution of goods. It is of urgent necessity that national and local governments in the OECD Member countries develop their efforts to reduce the adverse effects of motor traffic in urban areas. In so doing they should make provision for the needs of people who by choice or necessity do not have access to a car (e.g. children, the elderly, the handicapped), and should facilitate the safety and mobility of pedestrians and cyclists.

Extract from Conclusions of the OECD Conference, Paris, 1975 "Better Towns with Less Traffic"

NEW THINKING ABROAD

THE

The federal government is standing by with blueprints to ban all automobiles from the downtown areas of major American cities.

It is a political hot potato but experts say that air and noise pollution—plus the nightmarish traffic congestion — make such a drastic move necessary.

Norbert T. Tiemann, head of Federal Highway Administration, says urban traffic has reached the critical point.

"I personally feel that the time is not far off when we will have to bite the bullet and restrict private automobiles in the central business district, or at least a part of it, in many of our cities," Tiemann said.

THE SUNDAY TIMES, MARCH 14 19

Tyneside's tramway to beat t

Construction work began in November 1974 and so far 70 per cent of the north-south tunnels under Newcastle are finished; the whole system should be working by 1980—a 34-mile, 43-station light rail network, running from 6 am till midnight, and providing a train every two to seven minutes.

Planning director Howard finds it significant that the US Urban Mass Transit Administration, whose chief recently visited Tyne and Wear's test track at Benton, has recommended this kind of "semi-metro" as a "best buy" in urban transport. One virtue is that it is relatively cheap — the Tyne metro came out top in an economic comparison with an all-bus system, as well as far less harmful to the environment; another virtue is its flexibility.

The Tyne metro is comparatively cheap to build and run.

Building costs are kept down because it gets most of its 34 miles from existing British Rail routes, either closed or which would have been closed but for a £2½m-a-year Tyne and Wear

Solution: super-trams of a rapid transit system based on existing rail network

grant to keep them open pending the metro's completion. New route is limited to tunnelling under the centres of Newcastle and Gateshead, and two new surface stretches to bring the metro where it is needed at South Shields and Byker.

"We aimed at convenience and simplicity," says Howard. "We don't want to build monuments." He and his director - general, Desmond Fletcher, see the supertram as the backbone to an integrated public transport system, with connecting buses drawing up virtually alongside metro platforms; cheap and easy-to-use park-and-ride interchanges; and convenient interchanges with British Rail.

The cost has increased above the quoted when work be But officials say tha £3 million to cater people and discounti the current figure is per cent of estimate likely £200 million still sounds good pared with £25 milli urban motorway.

CAR-LESS CITY
OF THE FUTURE

a g athering of
rts in France
t his agency is
formation to assist
ies in their reluc-
inst traffic jams—
ulting racket and

cities to consider
notor traffic from
eas.

th the improved
ols, many areas of
remain heavily

bogged down with traffic,"
Tiemann said.

"The auto-free concept would
not only help solve this prob-
lem, it would also enhance the
environment a n d stimulate
downtown economies."

Thus far, the greatest attention has been focussed on
commercial areas with high
concentrations of pedestrian
traffic. But other areas could
benefit.

"Restricting cars from areas

of historic, or aesthetic importance would make them more
enjoyable for people."

Tiemann says car-free zones
should be a primary objective
in any comprehensive city plan.

"We are encouraging cities
to use federal aid highway
funds to implement s u c h
plans," he said.

Studies are also under way to
see if it would be possible to
"tax" away traffic. Vehicles
using congested streets during

peak traffic periods would be
charged.

A prime example cited by
Tiemann's staff is Minneapolis,
Minn., internationally known
for its solution to urban blight
and the ugliness of snarled
traffic.

More than 100 delegations of
planners from throughout the
world visit Minneapolis every
year to stroll along Nicollet
Mall and dream of doing the
same thing for their cities.

traffic
New book tells how to avoid
coronary and enjoy life

By Dr. David Nowlan,
Medical Correspondent

MORE pedestrian streets,
more bicycle paths on principal highways and a law pro-
hibiting the development of
new communities without
adequate recreational facilities for all age groups are
among the recommendations
in a new book on "Heart
Attack and Life Style" published yesterday by the Irish
Heart Foundation

Ramps laid down
to reduce speed

A new weapon against speeding
motorists goes into operation next
week for the first time on a public
road in Britain. The weapon: nine
humps to jog a motorist both
mentally and physically if he drives
too fast. Each hump is 12 ft. long
and 4 ins. wide.

The Government's Transport and
Road Research Laboratory is
installing the humps at Cuddesdon
Way in a housing estate on the
outskirts of Oxford. They are at
about 100 yard intervals.

The humps—nicknamed "sleeping
policemen"—are the first to appear
on a public road in Britain,
although they are in use in many
private roads. Other countries have
talled them on public roads.
P.A.)

69

MILJÖ- och HÄLSOVÅRDSFÖRVALTNINGEN
Telefon 69 05 00
Torkel Knutssonsgatan 20
Fack, 104 62 STOCKHOLM 17

STADSBYGGNADSKONTORET
TRAFIKPLANEAVDELNINGEN
Nils Ericsonsg. 17   411 03 Göteborg
Telefon 031 - 17 90 00

With reference to your letter about trafic policy I
can give you the following answer.

The trafic policy of Stockholm may briefly be described in

a) heavy transports are concentrated to a few certain
   thoroughfares and otherwise prohibited. Vehicles with
   a weight exceeding 3,5 ton are prohibited in the city
   between 10 pm and 06 am.

b) public transports are favoured with for example trafic-
   lanes for bus and taxi.

c) private cartrafic in the city is limited by few parking-
   possibilities and reduced passability across districts
   and sackstreets.

d) a program for trafic-clearance of all the town within
   5 years.

Our transport policy can briefly be described as
follows:

Firstly we want to increase the road safety for all
road users.
Secondly we want to increase the passability for
public transport and cyclists.
Thirdly we wish to keep the passabiblity for cars
on the same level as today. All the time we try
to reduce the environmental influence from the
traffic.

Yes we have cycleways. Since last year we have a
masterplan for the cycleway system (enclosed) and
we spend 2-4 000 000 Skr per year on the development
of this system.

We have several pedestrianised streets. You can see
them on the map of the enclosed parking information.

NEW THINKING ABROAD

STEVENAGE DEVELOPMENT CORPORATION
Daneshill House, Danestrete, Stevenage, Herts. SG1 1XD

ENGINEER'S DEPARTMENT   Telephone: Stevenage 4444
CHIEF ENGINEER
R. B. LENTHALL, C.Eng., F.I.MunE.

   In addition, there is a segregated cycleway system
running alongside all the primary roads and some of the secondary.
The cycleway system also offers segregation to pedestrians.
There is also a large area of the town containing about
30,000 people with a pedestrian segregated housing layout.

   In regard to public transport, the experimental
Superbus service has been in operation for 5 years. This
service is subsidised and at present deals with about one-third
of the town. The remainder is served by traditional bus
services which operate at comparatively poor frequencies
but are understood to make a small profit.

The practical experiences and experiments
reviewed at this Conference show that
policies combining selected traffic limitation
measures and public transport improvement
can achieve a better urban environment,
enhance accessibility for people and goods
and conserve energy. Therefore, the Conference concludes that the national and local
governments should actively support and
encourage the design, implementation and
evaluation of such programmes, through
their own political processes.

Extract from Conclusions of the OECD
Conference, Paris, 1975 "Better Towns with
Less Traffic"

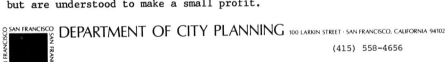

DEPARTMENT OF CITY PLANNING 100 LARKIN STREET · SAN FRANCISCO, CALIFORNIA 94102
(415) 558-4656

   The latest San Francisco Plan for Transportation was adopted as an
element of the City's Comprehensive Plan in 1972. It is notable for its
emphasis on the priority of public transit over automobiles for meeting
the transportation needs of residents and visitors.

   There are two major freeways in San Francisco, both in the same
general corridor, and current Comprehensive Plan policy calls for not
increasing (and reducing when possible) the vehicular capacity of bridges,
highways and freeways entering the city. It is unlikely that any addi-
tional freeways will be built in San Francisco.

ente Amster am
Stadhuis, O.Z. Voorburgwal

2. With regard to traffic-policy the city is divided in a central
   area-built prewar- and suburban areas (built post war). In the
   central area the policy is as follows:

   - no building of motorways;

   - reduction of motorvehicle traffic by parking-policy and traffic-
     free streets;

   - promotion of public transport by free tram- and buslanes and
     priorities on their behalf at the traffic-lights.

NICOLLET TRANSITWAY—MINNEAPOLIS The evolution of an auto-free zone.

NEW THINKING ABROAD

STEVENAGE
Segregated cycle- and pedestrian-ways linking Town Centre and residential and industrial areas and underpassing main-road junctions.

Traffic restraint and management

The Council has set itself a target to reduce the amount of traffic in central London by about one-third below the 1974 level. Following the results of public consultation, the Council has decided to seek powers to control the use of private non-residential car parks. For 1976-77, parking control will continue to be the principal method used.

Traffic restraint will be increasingly used to promote road safety, to give buses priority in traffic and to improve conditions for pedestrians. Lorries will be banned from designated areas. The current programme is to instal bus lanes at the rate of 25 a year but this figure will be increased. The Borough Councils will be encouraged and assisted to designate cycle ways on suitable minor roads.

Extract from Greater London Council's
Transport Policies and Programme for
1976-81.

The High Cost Of Our Road Building

From there, he went on to question Mr. Tully on the amount of money, in current and real terms, allocated for roads in 1974, 1975 and 1976.

The Minister provided the inflation rate for the relevant years, but not the real expenditure figures. Some £16 million had been allocated in nine months in 1974, he said, as against £21 million in 1975, and £20.4 million in 1976. To recover ground in the light of that statement, Mr. Tully spoke passionately about the increased commitment by the Exchequer to road building, rising, he said, from 50% of total expenditure in 1972/'73, to 58% this year. Furthermore, he added, the money was being allocated so as to afford maximum protection to the jobs of road workers.

Mr. Faulkner asked how Minister intended to employ people on less money. Tully responded that the work...

"increased committment to road-building"

58% of Local Government funds allocated to road-buildi

In the preceding pages one can see the attempts of other countries to solve the problems of urban traffic. The emphasis is on public transport, cycle-ways, pedestrianisation. In Dublin, contrary to the recommendations of the O.E.C.D. Conference in Paris in 1975 and the Declaration of Amsterdam, motorways and road-widening are the order of the day, public transport is down-graded to second place in our Development Plan, cycle-ways are non-existent, our one short stretch of pedestrianised street is not even paved across yet, and one of our two canals may be filled in to provide a motorway. The following sections discuss a number of straightforward alternatives open to us if we can persuade our planners to adopt them.

Transport Alternatives

The petrol engine is now recognised as one of the major causes of air pollution and general environmental destruction. A rapid deterioration in the quality of life, together with the changing world energy situation, has led to a widespread re-evaluation of various means of transport pushed out by the petrol engine. Traditional modes like the bicycle and the canal barge, and the improvement of rail transport offer many advantages over the systems which have replaced them, or are presently threatening their survival.

THE BICYCLE
In recognition of the obvious attractiveness of being quiet, clean, space-saving, health-giving and economic, while the motor-car is polluting, space-taking, expensive and frequently noisy, many cities in the United States, Britain and Europe have already provided or are planning suitable route networks to facilitate and encourage the use of the bicycle as a daily means of transport. Even before — and, in Ireland, without — official recognition of its advantages, there has been a remarkable growth in in bicycle popularity in recent years. In Ireland alone, the sale of bicycles has increased by 75% from 1972.

INLAND WATERWAYS
Recreational use of our inland waterways is growing. With increasing congestion on the roads, canals offer a number of advantages for the transport of freight. Standard barges can carry up to the equivalent of 20 lorries. By making full use of our canal system thousands of tons of freight could be removed from the city streets and much more from our national roads. What an obvious way of removing heavy port traffic from the city centre!

A recent innovation in the line of container traffic is barge bottom containers, which can be deposited directly into the water from the ship and floated to a destination which could be through the canals or up the Liffey to either of our main railway stations, Connolly or Heuston and then transported by rail.

RAIL TRANSPORT
In attempting to deal with the various dilemmas of urban travel, planners and policy-makers have been concerned almost exclusively with the private car. Even when public transport is considered, emphasis is given to roadways rather than railways.

The Dublin transportation study (D.T.S.) undertaken by An Foras Forbatha in 1970 follows this trend and shows a strong bias in favour of roads. The D.T.S. report did, however, assign a dominant role to public transport for city-centre access, and recommended that the suburban rail links be raised to their highest practicable level of public service, 'providing *all-day services and provided with feeder bus services* and expanded parking facilities at stations. Passenger service should be opened as far as Blanchardstown and Clondalkin *"before significant growth begins in these areas"*

Further recommendations were the restoration of the Harcourt Street Railway Line as a busway (why not as a railway?) and the investigation of a short underground in central Dublin 'that would more effectively connect up the four existing rail links to the city'. (The Architectural Review special issue on Dublin, November 1974 proposed a more modest underground line connecting Tara Street and Heuston Station, which could well prove more economical than the current C.I.E. proposal).

On implementing the full transportaiton plan and keeping up to date, the D.T.S. report recommends that *transport interests, community groups and the general public be consulted and that 'periodic reports should be issued reporting on the progress of transportation programmes'.*

Dublin's planning authorities have obviously ignored these important recommendations and are implementing their road-widening and highway programme regardless of public opinion, social and economic costs, and as if public transport in general and rail transport in particular should play only a very minor role in their planning strategy.

The only chink of light at present on the horizon is the Dublin Rail Rapid Transit Study published by Coras Iompair Eireann in April 1976 which would not have been included on the Amended Dublin Development Plan currently on exhibition, were it not for the pressure exerted by a number of councillors. Although the C.I.E. plan is a direct response to a strong recommendation in the D.T.S. Report, a large notice at the Development Plan Exhibition states that this is not part of the statutory development proposals for Dublin. The question to ask is *WHY NOT?*

PROPOSED RAPID TRANSIT SYSTEM FOR DUBLIN

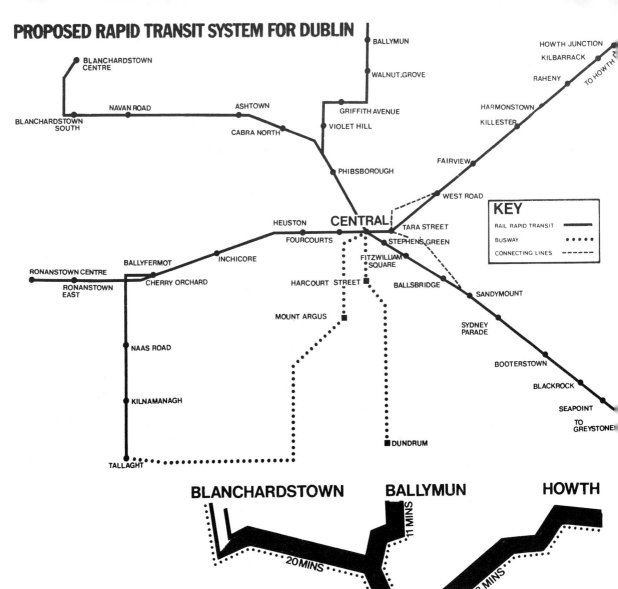

BLANCHARDSTOWN CENTRE
BLANCHARDSTOWN SOUTH
NAVAN ROAD
ASHTOWN
CABRA NORTH
BALLYMUN
WALNUT GROVE
GRIFFITH AVENUE
VIOLET HILL
PHIBSBOROUGH
HOWTH JUNCTION
KILBARRACK
RAHENY
TO HOWTH
HARMONSTOWN
KILLESTER
FAIRVIEW
WEST ROAD

KEY

RAIL RAPID TRANSIT	———
BUSWAY	• • • • •
CONNECTING LINES	- - - - -

HEUSTON
FOURCOURTS
CENTRAL
TARA STREET
STEPHENS GREEN
RONANSTOWN CENTRE
RONANSTOWN EAST
BALLYFERMOT
CHERRY ORCHARD
INCHICORE
FITZWILLIAM SQUARE
HARCOURT STREET
BALLSBRIDGE
SANDYMOUNT
MOUNT ARGUS
SYDNEY PARADE
NAAS ROAD
BOOTERSTOWN
KILNAMANAGH
BLACKROCK
SEAPOINT
TO GREYSTONE
DUNDRUM
TALLAGHT

THE C.I.E. PROPOSALS FOR PASSENGER MOVEMENT IN DUBLIN

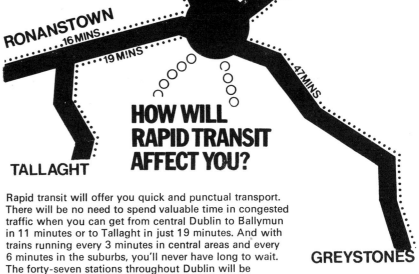

BLANCHARDSTOWN
BALLYMUN — 11 MINS
HOWTH — 23 MINS
20 MINS
RONANSTOWN — 16 MINS
19 MINS
TALLAGHT
GREYSTONES — 47 MINS

HOW WILL RAPID TRANSIT AFFECT YOU?

Rapid transit will offer you quick and punctual transport. There will be no need to spend valuable time in congested traffic when you can get from central Dublin to Ballymun in 11 minutes or to Tallaght in just 19 minutes. And with trains running every 3 minutes in central areas and every 6 minutes in the suburbs, you'll never have long to wait. The forty-seven stations throughout Dublin will be

74

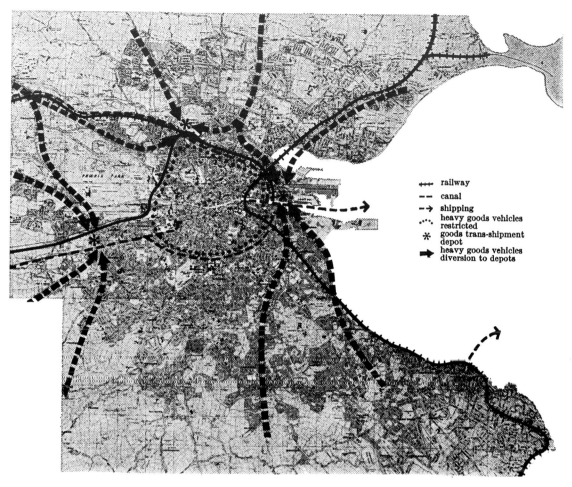

GOODS MOVEMENT: *The stars indicate very approximate positions of depots where loads are switched from heavy goods vehicles to smaller lorries and vans for distribution in the city. Heavy goods vehicles can themselves enter the city only at very slow speeds and by special permission. Diagram shows the canals brought back into use. Systems such as BACAT (barges aboard catamaran) allow barges to be transferred from sea going vessels straight into the canal system and vice versa. Diagram also assumes the modernisation of rail freight and the greater handling of goods by this medium.*

Reprinted from 'Architectural Review', November 1974.

ARCHITECTURAL REVIEW'S PROPOSALS for goods movement in Dublin.

Can Dublin be Saved?

It makes me wonder whether we are going backwards or forwards, to read in a guide book of Dublin written 65 years ago: *"No city has a more complete system of electric tramways . . .the cars run through Dublin and many of the suburbs at surprisingly low rates . . . cars run through about 50 miles of streets."*

Referring to railways the book notes: *"Most of the main line systems running into Dublin are connected by loop lines. The North Wall Station* (now closed) *the Amien Street Terminus, the Broadstone Terminus* (now Closed) *Kingsbridge Station, Westland Row Station, Harcourt Street Station* (now gone). *Frequent trams ran between them or near them all."*

So now, here we are — 65 years later — all our trams, several of our stations, and the city communities they once served, hanging on grimly by the skin of their teeth, in a constant battle against the ever tightening web of the road network with its resultant traffic blight, pollution and danger.

Can Dublin be saved? Yes, it can, despite the ravages it has already suffered if 'authority', in all its forms will stop planning things to suit its own way of thinking, and sit back and look at the city with the people who live in it and work in it, and see what we see, what we need, what we don't need, and what we can do to let Dublin breathe again.

Public participation is not simply letting the public look at plans which have been formulated by the Corporation. Public participation should be the participation of the public in the making of these plans, however unwieldy and difficult this would seem to be (and don't think it would be so unwieldy)in the long run much time and money, abuses and excuses, would be saved.

If the policy for the city was changed, the effect on the morale of the people would be enormous. The pressure and uncertainty hovering over the inhabitants would be lifted, plans could be made to reinstate for people (not traffic) the sad litany of streets, squares, corners etc. that have already suffered the disease of planning blight.

All this at a time when cities in other countries are dropping major road plans because of the enormous cost, both in money and in the effects they would have on the housing and environment in these cities, and because, in many cities where major road plans have been put into operation, they have generated so much extra traffic that they simply have not been effective. One begins to doubt the sanity of our planners.

Unfortunately, we have no say in the matter of appointing a Corporation official, but we certainly have a say in the election of a councillor. These road proposals traverse the entire city from the centre to the suburbs, and I am sure every councillor has some part of it in his area. Those who support the roads will have a hard time explaining his or her stand at the next election, when much of the road widening and traffic blight will have already started if the proposals are passed.

DO YOU AGREE WITH THE ELECTED REPRESENTATIVES?

The first line of defence in the protection of our city is our elected representatives, the councillors and T.D.s. While councillors can and should influence policy, it is up to the public to make them constantly aware of the problems at issue. The present councillors are:

Electoral Area 1

Alderman Vincent Manning (Non-Party)
Councillor Michael Cosgrave (F.G.)
Councillor Edward Brennan (F.F.)
Councillor Tom Duffy (Lab.)
Councillor Sean D. Dublin Bay Loftus (Non-Party)

Electoral Area 2

Alderman Paddy Belton T.D. (F.G.)
Councillor Paddy Dunne (Lab.)
Councillor Eugene Timmons T.D. (F.F.)
Councillor Tim Killeen (F.F.)
Councillor Mrs. Johanna Barlow (Non-Party)

Electoral Area 3

Alderman Jim Tunney T.D. (F.F.)
Councillor Luke Belton T.D. (F.G.)
Councillor Mrs. Alice Glenn (F.G.)
Councillor Billy Keegan (Lab.)
Councillor Danny Bell (F.F.)

Electoral Area 4

Alderman Dr. Hugh Byrne T.D. (F.G.)
Councillor Richard Gogan T.D. (F.F.)
Councillor Tom Leonard (F.F,)
Councillor Pat Carroll (Lab.)
Councillor Frank Sherwin (Non-Party)

Electoral Area 5

Alderman Kevin Byrne (Non-Party)
Councillor Michael Keating (F.G.)
Councillor William Cumiskey (Lab.)
Councillor Tom Stafford (F.F.)
Councillor Ray Fay (F.G.)

Electoral Area 6

Alderman Dr. John O'Connell T.D. (Lab.)
Councillor Patrick J. O'Mahony (Lab.)
Councillor Jim Mitchell (F.G.)
Councillor Lauri Corcoran (F.F.)
Councillor Mrs. Eileen Lemass (F.F.)

Electoral Area 7

Alderman Alexis Fitzgerald (F.G.)
Councillor Pat Cummins (F.F.)
Councillor Dan Browne (Lab.)
Councillor Fergus O'Brien T.D. (F.G.)
Councillor Brendan Lynch (Non-Party)

Electoral Area 8

Alderman Ben Briscoe T.D. (F.F.)
Councillor Michael Collins (Lab.)
Councillor James O'Keeffe (F.G.)
Councillor Sean Kelly (F.G.)
Councillor Joe Dowling T.D. (F.F.)

Electoral Area 9

Alderman Mrs. Carmencita Hedermann (non-Party)
Councillor Sean Moore T.D. (F.F.)
Councillor Ruairi Quinn (Lab)
Councillor Gerard Brady (F.F.)
Councillor Peter Kelly (F.G.)

Of this number only fourteen out of forty-five councillors were present when a decision was taken to pass the Draft Development Plan (Revision) for public display, complete with motorways and road-widening. This is an absolute disgrace, but it is up to the public to see that it does not happen again. If your councillor is not prepared to represent you at crucial meetings such as this, he or she is not worthy of your vote!

CORPORATION ROAD WIDENING PROPOSALS

SECTION A

Major Roads

1 New bridge across the River Liffey in the vicinity of Cardiff Lane on the south bank and Guild Street on the north bank.

2 Interchanges on each side of the river to connect the existing street system to the new bridge.

3 A new approach road to the bridge from the city boundary near Merrion Railway Gates.

4 A new approach road to the bridge from Clontarf Road at a point adjacent to the railway overbridge.

SECTION B

Approved Road Schemes being implemented at present

6 Dame Street from Palace Street to Exchange Court (C.P.O. confirmed).

7 Cuffe Street — from Harcourt Street to Wexford Street (C.P.O. confirmed for part of).

8 Lower Kevin Street — from Wexford Street to Bride Street New (C.P.O. being prepared).

9 Upper Kevin Street — from Bride Street New to New Street (C.P.O. being prepared).

10 Patrick Street — from Dean Street to Nicholas Street (C.P.O. being prepared).

13 Lower Bridge Street — from Cooke Street to Merchants Quay (C.P.O. being prepared).

17 Clanbrassil Street Lower — South Circular Road Junction (C.P.O. Public Inquiry held).

18 Ardee Street from Cork Street to The Coombe (C.P.O. being prepared).

19 Cork Street and Dolphin's Barn Street from Ardee Street to South Circular Road (C.P.O. being prepared).

20 Marrowbone Lane — Cork Street to Summer Street (C.P.O. being prepared).

21 St. John's Road/Con Colbert Road/South Circular Road Junction (property requisition to be negotiated).

23 Merrion Road — Sydney Parade to Nutley Lane (C.P.O. confirmed).

24 Donnybrook Road — from Eglinton Road to Brookvale Road (C.P.O. being prepared).

25 Sandford Road — Marlborough Road Junction (C.P.O. confirmed).

26 Milltown Road — Classon's Bridge to Milltown Bridge (C.P.O. being prepared).

27 Orwell Road Junctions with (a) Orwell Park; (b) Zion Road (C.P.O. being prepared).

28 Rathfarnham Village By-Pass including Butterfield Avenue Junction (tender received and under examination).

30 New Link Road from Templeogue Road to Terenure Road West (C.P.O. being prepared) — Sports Grounds.

31 Kimmage Cross to (a) No. 315 Kimmage Road Lr.

32 Drimnagh Road — from Walkinstown Road to Kildare Road (C.P.O. confirmed).

33 Bluebell Avenue — from Kylemore Road to No. 71, Bluebell Avenue (C.P.O. being prepared).

34 Bluebell Lane — from Camac Park to Old Naas Road (Property acquisition being negotiated).

35 Chapelizod By-Pass — from Con Colbert Road — Sarsfield Road Junction to city boundary (C.P.O. submitted to Minister).

37 Memorial Road Bridge (tender received and under examination).

39 Parnell Street — from Parnell Square West to Ryder's Row (C.P.O. being prepared).

40 North King Street from Ryder's Row to Bow Street (C.P.O. being prepared).

47 Howth Road at Castle Avenue (C.P.O. confirmed).

54 Richmond Road and Ballybough Bridge from No. 184, Richmond Road to Clonliffe Road/Ballybough Road Junction (C.P.O. being prepared).

56 Griffith Avenue Extension — East of Finglas Road (C.P.O. being prepared).

57 New Road from Finglas Road to Botanic Road — Hart's Corner (C.P.O. Public Inquiry held).

59 Old Finglas Road — from Finglas Road to Tolka Estate Road (Property acquisition being negotiated).

63 Ratoath Road — from 'Rose Lodge' near Dominican Convent to Griffith Avenue Extension including improvement of its junction with Griffith Avenue Extension (C.P.O. being prepared).

64 Navan Road — from Nephin Road to Ratoath Road (C.P.O. being prepared).

SECTION C

Road Schemes at present being planned but which are not so advanced as to be submitted to An Coisde Cuspoiri Coiteann yet. They will be considered in detail with An Coisde Cuspoiri Coiteann in due course.

5 Extension of Lombard Street East to City Quay.

11 New Road from Winetavern Street to High Street.

? Francis Street from Cornmarket to Iveagh Market.

14 Clanbrassil Street Lower — from Lombard Street West to Kevin Street Upper.

15 Charlemont Street — from Albert Place West to Charlemont Place.

16 Grand Canal Road — from Harold's Cross Bridge to Leeson Street Bridge, on north bank of canal.

22 Serpentine Avenue from Merrion Road to Railway Crossing.

29 New Road from City Boundary through Loreto Abbey to Nutgrove Avenue, and section of Nutgrove Avenue.

36 New Bridge near Sean Heuston Bridge.

38 Mary's Abbey from Arran Street East to Capel Street.

41 New Road connecting Benburb Street at Queen Street with Chancery Street at Church Street.

42 North Circular Road — Old Cabra Road junction (Hanlon's Corner).

43 Dungriffin Road, Howth.

50 Malahide Road — from end of dual carriageway to City Boundary.

51 Malahide Road — Elm Mount Avenue to Kilmore Road.

52 Malahide Road — Griffith Avenue to Clancarthy Road.

53 New Road from Clontarf Road to East Wall Road.

55 New Santry Bypass — from Collins Avenue to City Boundary.

58 Finglas Road — Old Finglas Road to Glasnevin Cemetery.

60 New Dual-Carriageway — Mellowes Road to Plunkett Avenue.

61 Broombridge Road — from Carnlough Road to Ballyboggan Road.

62 Ballyboggan Road — from Finglas Road to Ratoath Road.

65 Blackhorse Avenue — from McKee Barracks to City Boundary.

SECTION D

Long Term Roads Reservations - It is not proposed to initiate these Road Schemes within the next five years. The purpose of their inclusion in the Draft Variation of the Development Plan is to indicate to interested persons, Architects, etc., that it is intended to keep lines free from development and that if redevelopment takes place along these roads a set-back of building line may be required.

Major Roads

1 A new road with interchanges from the Royal Canal at Croke Park to the City Boundary at Santry.

2 A new road along the Royal Canal from Sheriff Street to the City Boundary at Ratoath Road.

3 A new road from a point on the Royal Canal west of Cross Guns Bridge to Finglas Road near Prospect Cemetery.

Other Roads

4 Sir John Rogerson's Quay from Creighton Street to Cardiff Lane.

5 City Quay from Moss Street to Creighton Street.

6 George's Quay — from Corn Exchange Place to Moss Street.

7 Moss Street from City Quay to Townsend Street.

8 Shaw Street from Townsend Street to Pearse Street.

9 Townsend Street from Hawkins Street to Moss Street.

10 Tara Street from Pearse Street to George's Quay.

11 Pearse Street from Westland Row to Tara Street.

12 Westland Row from Lincoln Place to Pearse Street.

13 Fenian Street from Lincoln Place to Denzille Lane.

14 Lincoln Place from Leinster Street South to Westland Row.

15 Nassau Street from Grafton Street to Dawson Street.

16 Grand Canal Street Lower from Albert Place to Clanwilliam Place.

17 New Road — continuation of Macken Street from Grand Canal Street Lower to Mount Street Lower.

18 Baggot Street Lower from Ely Place to Pembroke Street Lower.

19 Merrion Row from St. Stephen's Green to Ely Place.

20 Aston Quay from Bedford Row to Aston's Place.

21 Wellington Quay from Grattan Bridge to Asdill's Row.

22 Crampton Quay from Asdill's Row to Bedford Row.

23 Essex Quay from Fishamble Street to Grattan Bridge.

24 Wood Quay from Winetavern Street to Fishamble Street.

25 Merchants Quay from Bridge Street Lower to Winetavern Street.

26 Usher's Quay from Bridgefoot Street to Bridge Street Lower.

27 Usher's Island from Watling Street to Bridgefoot Street.

28 Parliament Street from Dame Street to Wellington Quay.

29 Temple Bar from Temple Lane South to Anglesea Street.

30 Essex Street East from Parliament Street to Temple Lane South.

31 Fownes Street Lower from Wellington Quay to Temple Bar.

32 Fownes Street Upper from Temple Bar to Dame Street.

33 Eustace Street from Wellington Quay to Dame Street.

34 Dame Street from Parliament Street to Anglesea Street.

35 Lord Edward Street at No. 9 Lord Edward Street.

36 Fishamble Street from Lord Edward Street to Essex Quay.

37 Christchurch Place from Nicholas Street to Werburgh Street.

38 Werburgh Street from Castle Street to Ship Street Little.

39 Bridgefoot Street from Thomas Street West to Usher's Quay.

40 Thomas Court from School Street to Thomas Street West.

41 Pimlico from The Coombe to Marrowbone Lane.

42 New Road — cotninuation of Cork Street from Ardee Street to New Street South.

43 New Road — linking King Street South with Longford Street Little.

44 Longford Street Little from Digges Lane to Aungier Street.

45 Longford Street Great from Aungier Street to Stephen's Street Upper.

49 Wexford Street from Montague Street to Cuffe Street.

50 New Road — linking Grantham Street and Hatch Street Upper.

51 Harcourt Road from Camden Street Upper to Harcourt Street.

52 New Road — Extension of Harcourt Street to Charlemont Street.

53 Richmond Street South from Harcourt Road to Albert Place.

54 New Road — from Canal to Clanbrassil Street Lower at Lombard Street West.

55 Donore Avenue from Cork Street to Parnell Bridge.

56 New Road — along north bank of Grand Canal from Clanbrassil Street Upper to Davitt Road.

57 Bow Lane West from Steevens' Lane to Kennedy's Villas.

58 South Circular Road from Bulfin Road to Conyngham Road.

59 Suir Road from Davitt Road to Bulfin Road.

60 Mount Brown from O'Quinn Avenue to Old Kilmainham.

61 Old Kilmainham from Mount Brown to South Circular Road.

62 Brookfield Road from Old Kilmainham to Cameron Square.

63 Dolphin's Barn from South Circular Road to Grand Canal.

64 Jamestown Road from Tyrconnell Road to end of Jamestown Road.

65 New Road — from end of Jamestown Road to Kylemore Road.

66 Le Fanu Road.

67 Killeen Road.

68 Grand Canal Street Upper from Clanwilliam Place to Haddington Road.

69 Shelbourne Road from Haddington Road to Ballsbridge.

70 Serpentine Avenue from Railway Level Crossing to Oaklands Park.

71 Anglesea Road from Simmonscourt Road to Anglesea Bridge.

72 Donnybrook Road from Brookvale Road to Belmont Avenue.

73 Beaver Row from Anglesea Bridge to City Boundary.

74 Merrion Road from Herbert Avenue to City Boundary.

75 Milltown Road from Sandford Road to Milltown Bridge.

76 Ranelagh Road from Charlemont Bridge to Mount Pleasant.

78 Highfield Road from Rathgar Road to Upper Rathmines Road.

79 Orwell Road from Highfield Road to Waldron's Bridge.

80 New Road — from Grand Canal to Lower Harold's Cross Road.

81 Cross Road at Harold's Cross Park.

82 Mount Argus Road.

83 Kimmage Road Lower.

84 Crumlin Road from Grand Canal to Kildare Road.

85 Custom House Quay and Memorial Road — junction improvement.

86 Bachelors Walk from O'Connell Street to Liffey Street Lower.

87 Ormond Quay Lower from Liffey Street Lower to Capel Street.

88 Arran Quay from Church Street to Queen Street.

89 Ellis Quay from Queen Street to Ellis Street.

90 Sarsfield Quay from Ellis Street to Liffey Street West.

91 Swift's Row from Ormond Quay Lower to Great Strand Street.

92 Abbey Street Upper from Capel Street to Liffey Street Lower.

93 Jervis Street from Great Strand Street to Parnell Street.

94 Summerhill from Gardiner Street to Portland Row.

95 Parnell Street from O'Connell Street to Gardiner Street.

96 Dominick Street Upper from Bolton Street to Western Way.

97 Father Matthew Bridge (Church Street).

98 Church Street from Inns Quay to North King Street.

99 Church Street Upper from North King Street to Constitution Hill.

100 Brunswick Street North from Church Street Upper to Morning Star Avenue.

101 Constitution Hill from Church Street Upper to Phibsborough Road.

102 Phibsborough Road from Constitution Hill to Connaught Street.

103 Mellowes Bridge (Queen Street).

104 Queen Street from Arran Quay to North King Street.

106 George's Lane from North King Street to Brunswick Street North.

107 Grangegorman Lower from Brunswick Street North to Kirwan Street.

108 Grangegorman Upper.

109 Stoneybatter from North King Street to Manor Street.

110 Manor Street from Stoneybatter to Prussia Street.

111 Prussia Street from Manor Street to North Circular Road.

112 Kitestown Road, Howth.

113 Balkill Road, Howth.

114 St. Fintan's Road/Strand Road Junction, Sutton.

115 Greenfield Road from Sutton Cross to Church Road, Sutton.

116 Howth Road from Clontarf Road to Bothar na Naomh.

117 Malahide Road from Fairview to Griffith Avenue.

118 Poplar Row from North Strand Road to Annesley Place.

119 Philipsburgh Avenue from Fairview Strand to Annadale Drive.

120 Richmond Road from Drumcondra Road to Convent Avenue.

121 Botanic Avenue from Drumcondra Road Lower to Mobhi Road.

122 Annamoe Road.

123 Annamoe Terrace.

124 North Circular Road from Aughrim Street to Prussia Street.

125 Old Cabra Road from North Circular Road to Cabra Road.

126 Blackhorse Avenue from North Circular Road to McKee Barracks.

127 Baggot Road from Blackhorse Avenue to Navan Road.

128 Martin's Row from Chapelizod Bridge to City Boundary.